GOD OF REALITY AND JUSTICE

Frederick John Miller

TABLE OF CONTENTS

BEGINNING

We have been given consciousness and thought. These gifts enable us to explore, learn, discover, analyze, and consider everything we observe. This includes our world, and it also includes everything we have learned about reality through every means of observation throughout history and to our present time.

Cosmology is the study of the origin and development of the universe. If you pick up a book on cosmology or browse it on the Web, it will explain that the universe began with the Big Bang. The Big Bang is the instantaneous bursting forth of the entire universe from a starting point of undefined location because space and time did not yet exist. Space and time became defined only as of the Big Bang, so the explanation goes.

So, a rational person will then say, how exactly did all that come about? Well, science claims to have this covered. The details of cosmic inflation say that there was nothing, just the void, and suddenly quantum fluctuations occurred in that nothingness, which is to say that nothingness is not absolute, that there's actually something there, even when there's nothing, a state of no matter and no energy. Double speak? Well maybe. Or, perhaps there was some form of energy, which we now know, according to Einstein, can be transformed into matter and potentially, back again. They've got it all figured out.

It sounds a little far-fetched, that suddenly everything that ever was and that ever will be burst forth out of nothing to give meaning to what we call time and space. Yet, from everything science has been able to determine, it's a fact. Actually, it's a theory since it can't be proven. Apparently, it just happened. That's all.

It actually sounds miraculous. It may not be scientific, but it's real. We're here, aren't we? At least, I think so. We have been given consciousness, the ability to know that we exist.

> "I think, I think I am, therefore I am, I think.
> Of course you are my bright little star,
> I've miles
> And miles
> Of files

Pretty files of your forefather's fruit
And now to suit our
Great computer,
You're magnetic ink.

I'm more than that, I know I am, at
least, I think I must be."
The Moody Blues, In the Beginning,
from On the Threshold of a Dream

Consciousness also seems miraculous but nonetheless real. As well, we have been given the ability to think, not only to think but to feel, see, hear, and to dream. All of these, give us the ability of observation. We can take all these observations and call them random happenstance, or we can see them as God's creation.

In the third century, the 1200s, Thomas Aquinas gave five logical arguments regarding the existence of God. One of the five was that there had to be a First Cause. You can't argue with that. Another was what he called the Final Cause, which is to say that God is the force that causes unintelligent objects to behave in predictable ways. Hmm. Let's think about that.

That's really what the laws of physics are all about, the idea that throughout the universe the laws of physics prevail, that there is consistency with regard to causation throughout the universe. Lately, cosmologists have theorized that this may not be true, that there may be exceptions to the laws of physics in some locations, in some circumstances, at some extreme level of greatness or smallness.

It may be true. It may not be. If it helps to resolve the paradoxes and inconsistencies that are observed, then it may be helpful. Science sometimes has to use fudge factors to explain the unexplainable. This is science's bridge too far.

For all his good intentions, Thomas Aquinas may have also overreached with his five proofs. Is it really possible to prove the existence of God? Isn't this a little presumptuous? God is big, and we are small. It is not for us to prove the existence of God. We may not be able to prove God, but we can discover God. God is always there. It's up to us to make the choice to let God into our lives as a force that can guide us and help us along.

We can, however, make observations of God through our observations of what is real. For one thing, evil is an obvious reality and has been throughout history. People wonder, if God exists, why does he allow evil to happen?

In World War I the soldiers on both sides experienced perhaps the worst that life has to offer, for as long as it lasted anyway. They lived in trenches full of mud and filthy water that would swallow men alive. They suffered from gangrene and dysentery from lice infestation. They were there to follow orders. To peer above the trench might mean an immediate shot through the head. The technology of war had evolved to the point where killing machines had been manufactured in mass quantities. Those men were under constant bombardment. They had to worry about mines being dug under their lines to be set off in such manner as to kill the most soldiers. They had to worry about mustard gas drifting downwind to their positions, gases that could blind and kill. And when ordered to "go over the top", they climbed up ladders to advance into "no man's land", where they could be mowed down by machine guns. They would encounter impassable fields of barbed wire covered by machine gun fire, all while under bombardment. If they did succeed in reaching the enemy trenches, they inflicted close fire and bayonet attacks on the enemy. If they fell in no man's land, they might be left there to die in agony or to sink into the mire.

How could God allow such a thing? Why would God allow such a thing? We see such evil in the world every day. Just watch the news. Our observations tell us that God does allow evil. This is where Aquinas's Final Cause comes into play. Evil happens because God created a universe of cause and effect. Cause and effect are built into reality. Good begets good, and evil begets evil. Just like the laws of physics, God was the beginning of cause and effect.

What about when bad things happen in life? Bad things happen to good people. It just seems to be bad luck. Someone became ill. Someone got in a car accident. Someone lost their job. We can try to be careful. We can try to prepare. We can learn from mistakes and become wiser. We can and should do everything we can to keep healthy and safe. Still things go wrong, sometimes desperately so. Wouldn't it be nice if in life only good things happened? That's certainly not the world we live in. That's not our reality. Our abilities to observe and to think tell us that these things happen.

Since original sin human lives have been compromised by sin:

"Cursed *is* the ground for your sake;

In toil you shall eat *of* it

All the days of your life.

Both thorns and thistles it shall bring

forth for you,

And you shall eat the herb of the field.

In the sweat of your face you shall eat

bread

Till you return to the ground,

For out of it you were taken;

For dust you *are,*

And to dust you shall return."

Genesis 3: 18 – 19

Life is hard and full of risk and danger. Because God is God of Reality, we can draw the conclusion that in God's world, such things do happen but that God is always there keeping an eye on things. There's nothing too big or too small for God's attention. He made subatomic particles, and he made super clusters of galaxies. Through God we gain wisdom to learn from the things that happen to us in this sometimes hostile world, and through God we gain the strength to endure. Prayer is the way to seek God's help with trials. We can pray for strength and wisdom as we seek God's help through trials. God can do miracles in answer to prayer. I've seen it happen in my own life, and I think many people can say the same.

SPACETIME

"In the beginning God created the heavens and the earth. The earth was without form, and void; and darkness was on the face of the deep. And the Spirit of God was hovering over the face of the waters."

Genesis 1: 1 - 2

What is the origin of the universe? Could there be multiple universes? Space and time would exist separately for each universe. If other universes exist, then it is impossible ever to know anything about them because space and time belong to a particular universe. The only universe we ever can know anything about is our own. Learning about our universe helps us to understand reality.

Our understanding of reality is relative to our perspective here on planet earth, a small planet of a medium size star, the sun. Photographed from the Voyager spacecraft, Carl Sagan called the earth, "a pale blue dot." The sun is one star among millions in our galaxy, The Milky Way. Our galaxy is one of millions in our local group of galaxies, and these are part of clusters of galaxies. Clusters are part of superclusters, and the universe is full of superclusters of galaxies.

Our universe includes stars and all kinds of other celestial objects that are receding from each other, and from us here on earth at accelerating rates. This is determined from spectra of light from stars and other celestial objects that reach us here on earth. We determine distances by means of supernovae that have relatively consistent light intensity. These are 'standard candles' of known luminosity.

Knowing the distance to celestial objects makes it possible to know how fast they are moving away from us. The light spectrum of an object is red-shifted as the distance increases. This is due to the Doppler effect. The wavelength of light is shorter for objects moving toward us and longer for objects moving away from us, such as is heard when a police car with its siren on is approaching or receding. Receding objects are said to be red-shifted.

The Hubble constant is an estimated measure of whether the universe will continue to accelerate apart, start to come back together, or stay the same. Observation and calculation show that the universe seems to be flying apart faster and faster.

Einstein's law of special relativity establishes that light travels at approximately 3×10^8 meters per second. Light equates to mass according to Einstein's equation, $e = mc^2$, where e is energy (in photons) and m is mass, and c is the speed of light. Nothing can travel faster than the speed of light. Yet there are objects receding from all other objects faster than the speed of light because space itself is expanding as it has since its beginning.

Einstein's theory of general relativity established that gravity actually bends or curves space so that objects moving in freefall through space follow a curved path depending on the gravity from massive heavenly objects, such as black holes. This is true even with respect to light. In fact, objects whose location would otherwise be hidden behind other large objects are actually visible due to gravitational lensing, which is curving of light paths due to gravity.

The idea that space could be shaped in this way is counter-intuitive, but the best example is that of a ball being released with horizontal force along the lip of large funnel. The ball would roll around the inside of the funnel and gradually continue to roll around lower and lower until it falls into the hole at the bottom. If an object were so massive it could never escape from the hole at the bottom of the funnel, that would illustrate a black hole.

If an object were moving sufficiently fast, it would be drawn steadily toward the black hole, but then it would continue on its way with its path altered. This is just as if a space vehicle such as the Voyager spacecraft used the gravity of Saturn to build up sufficient speed to cause it to be deflected with a boost from the gravity of Saturn to send it off to other planets and even out of the solar system into interstellar space.

In golf a putt that rims the cup is moving too fast to fall in but rather continues along an altered path. This too is described as curved space.

In the funnel example, if the ball were not moving fast enough once set in horizontal motion, or if the object at the bottom of the funnel were sufficiently massive relative to that speed, then the ball would reach the bottom of the funnel and could never escape that hole. In space this is known as a black hole.

A black hole is the remnant of a very large star that has exhausted its fuel, mainly hydrogen and helium. If a star is sufficiently large it will ultimately collapse on itself. It will become so dense that the entire mass of the star will become so packed and its gravity so great that nothing, not even light, can escape.

A black hole was first theorized to exist even though it couldn't be seen. It was predicted based on the apparent gravitational pull it exerted on other celestial objects relatively nearby.

The universe has been traced back to what has been called "the Big Bang". The Big Bang was an instantaneous inflation from a single point in space at a time when space did not yet exist. According to the idea of expanding space, this single point of space was the only space at that time, and the whole universe began at that one and only point of empty space.

The Big Bang had to have some kind of initiation. Science explains that inflation, the process of instantaneously expanding to create the universe, came about through what is called quantum fluctuations in empty space.

Because spacetime was undefined prior to cosmic inflation, the beginning of the universe came about through characteristics of empty space. There is something called zero-point energy. Apparently, empty space is not really empty. Instead, there is vacuum energy which, if it is non-zero, causes empty space to actually consist of vacuum energy. A vacuum that is free of mass is not necessarily free of energy. This is evidenced by the fact that energy can pass through empty space.

Think about how light passes through the empty space between the sun and the earth. Even more amazing is that light from stars billions of miles away can reach us. Light travels at 3×10^8 meters per second. In one second light can travel 3×10^8 meters. That's 3×10^5 kilometers. In miles that's about 1.6×10^5 miles. We can now see light from stars that shined millions of years ago and which may no longer exist at all or may have become black holes or neutron stars. As the stars recede from us, due to expanding space, we can effectively see back in time. With the help of the Hubble telescope, we can see back to about 13 billion years ago. The Big Bang is estimated to have been about 14 billion years ago. The Hubble telescope orbits the earth and is currently the most powerful telescope. In the coming decade the James Webb telescope will extend our ability to see farther back in time than the Hubble.

Light travels through millions of miles of empty space. Gravity also travels through millions of miles of empty space. Gravity keeps the planets in orbit around the sun. Gravity holds the galaxies together. In effect, although expanding space applies to objects within galaxies, those objects within galaxies are not flying apart relative to the rest of the galaxy. Rather the galaxy itself is flying apart from other galaxies at expanding rates due to expanding space.

When light reaches an opaque object, it may be reflected. In the case of a transparent or translucent object, light will pass through it. If it passes through a prism, white light produces a spectrum through refraction due to differences in the thickness of the prism according to the differences of the wavelengths and frequencies of the various colors.

Visible light is electromagnetic energy, which ranges in wavelength and frequency from gamma radiation to radio waves. In order from highest frequency and shortest wavelength to lowest frequency and longest wavelength the types of electromagnetic radiation are: gamma radiation, xrays, ultra violet, visible light, infrared, microwaves, and radio waves.

Light and all electromagnetic radiation travel through gases, such as the earth's atmosphere as well as through empty space, such as the space between the sun and the earth.

Although it passes through anything including planets and stars, gravity is much weaker than electromagnetic energy but extends out into space without limit. Like electromagnetic energy, it diminishes according to the inverse square law, which states that the intensity decreases proportionally to the square of the distance. Like electromagnetic energy, gravity passes through empty space. Gravity keeps the planets in orbit around the sun, and it even holds galaxies together.

At the center of most galaxies is a black hole that originated from extremely massive stars that collapsed on themselves and became so dense that nothing could escape from it, not even light. Hawking radiation is emitted as a byproduct of a black hole, and only in this way do black holes dissipate whatsoever.

The fact that both light and gravity can pass through empty space is remarkable. How exactly does that happen? The answer lies in vacuum energy. "Empty space" is not necessarily completely empty. Even if there is no matter at all present, there is energy. Zero-point energy is the lowest energy vacuum there could possibly be. It is unknown whether zero-point energy exists anywhere, and it is unclear whether the space around us is zero-point energy. If not, then the theoretical possibility exists that the entire universe could pass into a lower energy state, perhaps still greater than zero-point energy. Empty space is the key to understanding the universe. To understand why, it's necessary to understand what things are made of.

There is something called, "the Standard Model", and it identifies what matter and energy actually consist of. The Standard Model identifies the particles that make matter as quarks. There are six of them, and they have been given the names: up, down, strange, charm, top, and bottom. Quarks have a property of spin, which for quarks is one-half. Other particles that make up matter are electrons, muons, and neutrinos.

The Standard Model also identifies force carriers of energy, which are gluons, photons, bosons, and possibly gravitons. The existence of gravitons is predicted but has not been proven. Of these, only photons convey electromagnetic radiation including light, but photons carry all electromagnetic radiation from gamma rays to radio waves.

One boson has spin equal zero, and it plays the key role in making matter at the particle level. In Einstein's Special Relativity matter and energy equate to each other and can transition from matter to energy or from energy to matter according to $e = mc^2$ where e is energy, m is mass, and c is the velocity of light. The Higgs boson converts energy to matter, and the main thing that makes matter is that with matter no two quarks can occupy the same position in space at the same time.

Energy, in the form of photons, such as light, can occupy the same space at the same time as another photon and can pass through each other. Photons have integer spin, and quarks have half spin. Higgs bosons have zero spin.

Quarks form fermions, which are particles that make up matter consisting of protons, neutrons, and electrons, and these combine to make atoms, which in turn combine to make molecules, which make up all the elements in the periodic table of the elements.

According to $e = mc^2$, light can become matter and matter can become light. In atomic fission, such as occurs in a nuclear reactor or in an atomic bomb, atoms are split in a controlled (reactor) or uncontrolled (bomb) manner. Atomic fission produces a huge amount of energy in the form of light, heat, and gamma radiation. Matter and energy are different forms of the same thing and one can become the other at the particle level.

Under high enough energy photons can become matter in the form of particles and anti-particles. Antiparticles are identical to particles of matter except that they have reversed charge. Matter consists of the proton and antiproton as well as the electron and the positron. When particles of matter collide with particles of antimatter, they release energy in the form of a photon.

Antimatter consists of particles created at the Big Bang, with one antiparticle created for each particle of matter. Particles of antimatter have the opposite charge of the corresponding particles of matter. Thus, a proton has a positive electric charge, and an antiproton has a negative charge. An electron has a negative charge, and a positron (anti-electron) has a positive charge. One unexplained mystery of the universe and of physics is that, if the creation of a particle of matter is accompanied by the creation of a particle of antimatter, then where did all the antimatter go?

The Hubble constant is a number relative to zero, which is based on all the gravity of all the matter and energy in the entire universe. This is determined by the observable gravity based on the average density of the universe. Gravity holds the galaxies together and holds clusters of galaxies and superclusters of galaxies together.

Cosmic Microwave Background (CMB) consists of remnants of the Big Bang. Study of the CMB indicates that the universe is expanding at an accelerating rate, and this is the same conclusion as is determined from the observed gravity in the universe.

The observed gravity in the universe is much greater than can be explained by the observable mass. Scientists believe there is mass called dark matter, that makes up 25% of the universe, and that there is energy known as dark energy, which makes up 70% of the universe. There has been no known observation of either dark matter or dark energy. Scientists conclude that they must exist in order to explain the observed gravity that holds the galaxies together (dark matter) and to explain the apparent expansion of the universe (dark energy). Dark energy appears to work opposite gravity and to be the main cause of the expansion of the universe. Thus, 95% of the universe consists of dark matter and dark energy, and only 5% consists of the matter that we're familiar with.

Light has both a wave and a particle nature. Einstein discovered the photoelectric effect by showing that light of sufficiently high energy could dislodge an electron. The energy required is proportional to the frequency of the light. Light also has a wave nature, which is more intuitive. The Michelson experiment put a single beam of light through two slits to make an alternating fringe pattern, showing that light exhibits reinforcing and canceling wave characteristics. This is known as the wave-particle duality of light.

The photoelectric effect showed not only that light had a particle nature but also that the intensity (frequency) of the light was the sole determinant of the energy necessary and that that energy increased in quanta, or discrete packets.

Quantum theory tells us that there are minimum discrete packages of natural basic building blocks, such as light, that limit how small a photon of light can possibly be. Quantum mechanics tells us that these are minimum-sized packets of light in terms of both wave and particle nature that make light the unique energy that it is. These properties are actually common to the entire range of the electromagnetic spectrum. Quantum mechanics seems at first counter-intuitive but is actually understandable taking into consideration that light and its properties are observed with light, traveling at the speed of light.

INTERSTICE

The challenge of a lifetime is to learn and understand truth. The universe exists. Everything that ever happened had a cause. We explore causation backward and forward in time. There must be an initial cause of cosmic inflation. That cause is God. God is God of all, including any and all universes. The Bible is the source of written information about God and his relationship with living, conscious humankind. God is the God of all reality. Belief in God means that everything from the creation of the universe to human life and consciousness is due to the one ultimate reality, and that is God.

It is science that shows us how much we don't know about the origin of the universe. We see and sense that which seems consistent with the everyday facts of science. Isaac Newton understood the world and its laws to be readily observable. Einstein showed us that reality is far more complex than Newton ever realized.

God is God of reality. In the quantum fluctuations that occur in vacuum energy God created space and time as well as the matter and energy. Energy consists of electromagnetic radiation, gravity, the strong force, and the weak force. Electromagnetic radiation includes the entire electromagnetic spectrum ranging from gamma radiation to infrared. Gravity holds the universe together by attracting both matter and energy. The strong force holds the nuclei of atoms together, and the weak force controls nuclear decay, which is the emission of neutrons to form isotopes.

All forms of energy act through the boson, which is a fundamental particle, with integer spin as opposed to fermions, which have one-half spin. Unlike fermions, such as neutrons, protons, and electrons, bosons can pass through each other, since they are not matter, which was formed through the interaction of the Higgs boson. This property of bosons, that they occupy no space and can exist in the same space as other bosons, is why light can pass through light years of space through gases or through a vacuum, the relative absence of matter or antimatter.

Bosons are also the means by which gravity acts through light years of space, whether it be matter or empty space. Gravity passes through planets and stars and acts to attract other matter on the other side of those objects.

Fermions and bosons and the quantum
fluctuations that occur in the zero-point
energy of a vacuum are the means by which
God created the universe, and they are the
means by which the universe evolves
through space and time. As the universe
goes forward in spacetime, every living
thing and every conscious being lives its
life.

The beginning of the universe is not
explained by quantum fluctuations that
occurred amid nothingness. The scientific
theories of creation keep going back one
step more. They never identify an original
cause but rather start the process one step
earlier, leaving original causation undefined.
Ultimately, there is no explanation for an
original cause. We are left wondering how
quantum fluctuations could occur when

space and time had no meaning. Any possible explanation would only lead to questions about how that theorized occurrence could have been caused.

Instead of explaining the origin of the universe, those theories actually say that there had to be a cause for creation to have happened. That original cause has been named the Big Bang, which itself had to have an original state and cause. An analogy is that science has theorized that there must be "dark energy" because the universe has been observed to be ever-expanding and ever accelerating into newly expanded space. Or, for example, "dark matter" has been proposed as the explanation for the observation that there apparently is more gravity needed to hold the galaxies together than can be explained by all the observed matter in the universe.

Neither dark energy nor dark matter have ever been observed, and yet there is near agreement that they both must exist. To say that they must exist is really to say that they are the names that have been assigned to this otherwise unexplained need to identify causation for the anomaly.

In both cases, that of dark energy and that of dark matter, science is actually saying that these are theorized explanations for the observations that have been made. This is much different from discoveries that can be demonstrated and are repeatable because they are real. They have predictable effects that can be used to solve real problems and provide real solutions, such as all the math and physics that was effectively proven when man successfully reached the moon and safely returned. Further evidence are the successful launches of the Mariner I and II and the Voyager I and II spacecraft out of our entire solar system and into interstellar space.

Theories that cannot be directly observed and that cannot be demonstrated or repeated are different. They are merely unproven theories that have been proposed to solve problems for which no solutions have been identified or explained with certainty.

In every case, causation which cannot be explained with the same kind of certainty that we have for the math and physics that put a man on the moon could also be identified as the work of God, the means by which God created and maintains the universe including us and our world.

JUSTICE

No matter how far we look into the distant past or into the distant future, there is no explanation that works except that it started somehow and 'somehow' can only be God. The fact that the universe exists at all means God is real and is working in everything that happens everywhere every day.

This is not to say that God is "only" the God of the galaxies. It is to say that because God is "God of the galaxies", he is also the God of everything throughout all space and time including you and me and all of us. Because God is real and because He is God of everything in space and time, everything He wills for our lives, every value that he has taught us has the authority of God.

If we realize that God is the God of all reality, then the values He has given us and the teachings He has taught us have the ultimate authority, that of God. This means that values matter. The Old Testament provides ample evidence of the wrath of God, reason to live life with the fear of God, to live our lives in ways that show respect and obedience to God. The rules we live by are not random. They are the rules He has given us, and those rules prevail over us and over all else because they have been given to us by the God of reality.

The earliest evidence of God's expectations for our lives is the Great Flood:

> "Then the Lord saw that the wickedness of man was great in the earth, and that every intent of the thoughts of his heart was only evil continually. And the Lord was sorry

that He had made man on the earth, and He was grieved in His heart. So the Lord said, 'I will destroy man whom I have created from the face of the earth, both man and beast, creeping thing, and birds of the air, for I am sorry that I have made them.'" But Noah found grace in the eyes of the Lord."

Genesis 6: 5 – 8

Then on Mount Sinai God gave Moses the Ten Commandments:

I am the Lord your God

You shall have no other Gods before me

You shall not make for yourself any carved images, any likeness of anything that is in heaven above, or that is in the earth beneath, or that is in the water under the earth

You shall not bow down to them or serve them.

You shall not take the name of the Lord your God in vain.

Remember the Sabbath day, to keep it holy.

Honor your father and your mother.

You shall not commit murder.

You shall not commit adultery.

You shall not steal.

You shall not bear false witness against your neighbor.

You shall not covet."

Exodus 20: 2-17

God's plan for us is to live together in peace:

"He shall judge between the nations,
And rebuke many people,
They shall beat their swords into plowshares,

And their spears into pruning hooks,
Nation shall not lift up sword against
nation,
Neither shall they learn war
anymore."
Isaiah 2: 4

The peace that God intends for us depends
on putting God first. God must be in charge
of our lives. To do otherwise is to invoke
the wrath of God.

"Therefore, you shall keep the
commandments of the Lord your
God, to walk in his ways and to fear
Him."
Deuteronomy 8: 6

Since the time of creation there had been no
written Bible. Yet by the time of Noah sin
had become so pervasive that God wiped out
all except for Noah and his family, who after

the flood multiplied and repopulated the earth. From the time of Adam and Eve in the garden of Eden, man and woman learned the consequences of sin.

"And out of the ground the Lord God made every tree grow that is pleasant to the sight and good for food. The tree of life was also in the midst of the garden, and the tree of the knowledge of good and evil."
Genesis 2:9

"Then the serpent said to the woman, 'You will not surely die. For God knows that in the day you eat of it your eyes will be opened, and you will be like God."
Genesis 3: 4 – 5

"Then the Lord God said, 'Behold the man has become like one of Us, to know good and evil. And now, lest he put out his hand and take also of the tree of life, and eat, and live forever' – therefore the Lord God sent him out of the garden of Eden to till the ground from which he was taken."

Genesis 3: 22 – 23

Ever since the Garden of Eden, people have lived in sin. Because of the reality of God, there is right and wrong, and what you stand for matters. What you stand for is the sum total of your life, which demonstrates to God whether you've lived your life in love and fear of God. Whether or not that is so is the statement we make before God, and that statement is the basis on which each of us will ultimately be judged before God.

With the Ten Commandments God gave us explicit rules to live by. In a lifetime of challenges, imperfect man and woman try to obey God's commandments. Each time we fail to do so, we commit sin in the eyes of God.

To commit sin is to break God's law. It is important to distinguish between God's law and man's law. There are many things that we are told we should do. These dictates and obligations come from man. God's law comes from God, and it is the sense of right and wrong that comes from God and which we know in our hearts. To obey God's law is not necessarily what people want us to do.

Those that seek to obey God's law do so imperfectly. In the real world choices are sometimes difficult and complicated. Ever since original sin in the garden of Eden, men and women have struggled with sin, due in part to the demands of the world and in part to the sinfulness of man.

> For the good that I will *to do,* I do not do; but the evil I will not *to do,* that I practice. Now if I do what I will not *to do,* it is no longer I who do it, but sin that dwells in me. Romans 7: 19 – 20

Obedience to God's law is the demonstration of justice. We find justice within ourselves by recognizing our sins and demonstrating a sincere willingness to change our hearts from within.

> "And He said, 'What comes out of a man, that defiles a man. For from within, out of the heart of men, proceed evil thoughts, adulteries,

fornications, murders, thefts, covetousness, wickedness, deceit, lewdness, an evil eye, blasphemy, pride, foolishness. All these evil things come from within and defile a man.'"

Mark 7" 20 – 22

"For a good tree does not bear bad fruit, nor does a bad tree bear good fruit. For every tree is known by its own fruit. For men do not gather figs from thorns, nor do they gather grapes from a bramble bush. A good man out of the good treasure of his heart brings forth good, and an evil man out of the evil treasure of his heart brings forth evil. For out of the abundance of the heart his mouth speaks."

Luke 7: 43 - 45

Because of the reality of God, we are inferior to God. What we stand for demonstrates whether that truth has guided our lives, whether we love and fear God. God's gifts to us include consciousness and thought. These abilities have made possible all that science has discovered and accomplished over the centuries.

By giving man consciousness and thought, God has given us the ability to make discoveries and to achieve great things. But consciousness and thought also give us the potential for humility before God. They are the means by which we can glimpse the truth, that we are created by God and that we exist at all only through the grace of God.

"Surely you have things turned around!

Shall the potter be esteemed as the
clay;
For shall the thing made say of him
who made it,
'He did not make me'?
Or shall the thing formed say of him
who formed it,
'He has no understanding'"?
Isaiah 29: 16

What we stand for is demonstrated by
whether we treat others the way we'd want
to be treated ourselves. This is the Golden
Rule, and it is the pillar of justice.

"And just as you want men to do to
you, you also do to them likewise."
Luke 6:31

> "Therefore, whatever you want men
> to do to you, do also to them, for this
> is the Law and the Prophets."
> Matthew 7: 12

The elite who have denied justice will be
brought down.

> "So the last will be first, and the first
> last. For many are called, but few
> chosen."
> Matthew 20: 16

To believe in God is to believe not only in
theory but in deed. Many who are moral
and believe in justice are repelled by the
hypocritical behavior of those who espouse
their own self-serving points of view in the
name of religion.

"But be doers of the word, and not hearers only, deceiving yourselves. For if anyone is a hearer of the word and not a doer, he is like a man observing his natural face in a mirror; for he observes himself, goes away, and immediately forgets what kind of man he was. But he who looks into the perfect law of liberty and continues *in it,* and is not a forgetful hearer but a doer of the work, this one will be blessed in what he does."

James 1: 22 – 25

As an example, to see the risks taken and sacrifices made by our healthcare workers during the pandemic and to honor them with words and celebrations is fine, but to live our faith honestly is to do what we can to support them in meaningful ways: seeing that they have the protective equipment they need in a timely manner, giving them the flexibility needed to perform their service while maintaining their own health and safety and that of their families, and to pay them commensurate to their service and sacrifice.

There is no better example of living one's faith than the story of the good Samaritan:

> "So he went to *him* and bandaged his wounds, pouring on oil and wine; and he set him on his own animal, brought him to an inn, and took care of him. On the next day, when he departed, he took out two denarii, gave *them* to the innkeeper, and said to him, 'Take care of him; and whatever more you spend, when I come again, I will repay you.' So

which of these three do you think was neighbor to him who fell among the thieves?"

And he said, "He who showed mercy on him."

Luke 10: 34 - 37

Isaiah prophesied that God would establish justice, not man's justice, but God's justice:

"Of the increase of His government and peace
There will be no end.
Upon the throne of David and over His kingdom,
To order it and establish it with judgment and justice
From that time forward, even forever.
The zeal of the Lord of hosts will perform this."

Isaiah 9: 7

John the Baptist warned of the consequences
of sin and hypocrisy, that we should live
lives worthy of repentance:

> "Brood of vipers! Who warned you
> to flee from the wrath to come?
> Therefore, bear fruits worthy of
> repentance.... And even now the ax
> is laid to root of the trees. Therefore
> every tree which does not bear good
> fruit is cut down and thrown into the
> fire.
> Luke 3: 7 – 9

In the end there will be justice:

> "At that time Michael shall stand up,
> The great prince who stands watch
> over the sons of your people
> And there shall be a time of trouble,

Such as never was since there was a
nation,
Even to that time.
And at that time your people shall be
delivered,
Every one who is found written in
the book.
And many of those who sleep in the
dust of the earth shall awake,
Some to everlasting life,
Some to shame and everlasting
contempt.
Those who are wise shall shine
Like the brightness of the firmament,
And those who turn many to
righteousness
Like the stars forever and ever."
Daniel 12: 1 - 3

The universe will end as it began, with the
miraculous power that only God can wield.

The universe will continue to expand, and all the heavens will recede. Again the impossibilities will be resolved. All matter will become energy, and that energy will gradually over eons dissipate until there is only the void. The elementary particles that make up all matter will return to photons per $e = mc^2$. Energy and matter will fade away until there is only the absolute vacuum of zero-point energy. Everything that ever has been will again become void as it was in the beginning. Even then God will say, "I AM." as he once told Moses in the book of Exodus. Perhaps, He will again deem to create a new instance of the universe, life, humankind, and consciousness.

God's justice will have been served. There will be only the Spirit of God. He will watch over the souls of the redeemed.

"And I heard a loud voice from heaven saying, 'Behold the tabernacle of God is with men, and He will dwell with them, and they shall be his people. God Himself will be with them and be their God. And God will wipe away every tear from their eyes; There shall be no more death, nor sorrow, nor crying. There shall be no more pain, for the former things have passed away."
Revelation 21: 3 – 4

"And He said to me, 'It is done! I am the Alpha and the Omega, the Beginning and the End. I will give of the fountain of the water of life freely to him who thirsts. He who overcomes shall inherit all things, and I will be his God, and he shall be My son. But the cowardly,

unbelieving, abominable, murderers,
sexually immoral, sorcerers,
idolaters, and all liars shall have their
part in the lake which burns with fire
and brimstone, which is the second
death.'"

Revelation 21: 6 – 8

"He who is unjust, let him be unjust
still; he who is filthy, let him be
filthy still; he who is righteous, let
him be righteous still; he who is
holy, let him be holy still."

Revelation 22: 14 – 15

For all science has shown us it can not avoid
being based on impossibilities. Without
God it is impossible for this entire vast
universe to have burst from a single point of
zero-point energy, a vacuum; it is
impossible for everything that is to have

always been; it is impossible for everything that is to turn to nothing. It is at both end nodes of time and space that we see the reality of God.

It is not only at the end nodes that we see the reality of God. We see the reality of God in the beauty and goodness that we find in the world and in each other.

All that we have learned from science tells us what it is that we don't know. It is science that shows the reality of God. Because of the reality of God, there is right and wrong. To believe in God is to accept that what we stand for matters.

For all the achievements of science, science can never provide a thorough explanation of creation. Science can never thoroughly explain the initiation of the creation process no matter how many steps in the process it claims to have identified. Neither can it thoroughly explain the ultimate end of time and space. It is just as impossible for nothing to become everything as it is for everything to become nothing.

The explanations are incomplete because there can be no thorough explanation for the miraculous, a concept unknown to science. I believe in God because I believe He is truth. Because God is Truth, we have values. And because we have values, there is right and wrong. In a world where science holds credibility, and matters of the spirit are often relegated to the non-humanist realm, there's a point where science gets stuck and starts to conceive alternative, unproveable explanations, which no longer have the same degree of credibility as Newtonian physics, for example.

Learning science with an open mind opens the door to the reality of God.

Consciousness and thought can help us find true humility. Yes, the Bible was conveyed and written by men and women, but we would not have it if God had not intended to speak through it over the centuries. God commands us to believe that HE IS, to worship Him and fear him. He gives us hope for a life that can't be defeated by death on earth. If there's hope, then we should hang onto that hope and get on board.

> "Though I've grown old, the bell still rings for me, as it does for all who truly believe."
> The Polar Express

Acknowledgements

First and foremost, I thank Marilyn Buckley for proof-reading and editing including grammatical improvements and rewording. She also suggested information and ideas related to her extensive background in religion, particularly Christianity.

I'd also like to thank Rev. Dennis Tebeest, whose sermons woke me up to the fact that God is not only "God of the Galaxies" but is ever-present and in touch with our everyday lives.

I'd like to thank William Mackay Miller, who set an example that encouraged me to write and who taught me to look for interstices because, "that's where the action is".

This book comes from a science education and a life long interest in what is real, what is there to find out using my mind, my ability to observe, to learn, and to think. I spent years researching physics and cosmology, however daunting, and I complemented that by reading the Bible through five more times consecutively.

I've had the feeling for some time that the origin of the universe is more mysterious than science can fully explain, and I found that that fact has profound consequences for discovering God and His mastery of Reality and Justice.